Maths Revision Booklet
for CCEA GCSE 2-tier specification
M3

Neill Hamilton

Name:

© Neill Hamilton and Colourpoint Creative Ltd 2019

ISBN: 978 1 78073 194 0

First Edition
Third Impression 2024

Layout and design: April Sky Design
Printed by: Peninsula Print and Design

All rights reserved. No part of this publication may be reproduced, stored in a retrieval system or transmitted in any form or by any means, electronic, mechanical, photocopying, scanning, recording or otherwise, without the prior written permission of the copyright owners and publisher of this book.

Colourpoint Educational
An imprint of Colourpoint Creative Ltd
Colourpoint House
Jubilee Business Park
21 Jubilee Road
Newtownards
County Down
Northern Ireland
BT23 4YH

Tel: 028 9182 0505
E-mail: sales@colourpoint.co.uk
Web site: www.colourpoint.co.uk

Publisher's Note: This book has been written to help students preparing for the GCSE Mathematics specification from CCEA. While Colourpoint Educational and the authors have taken every care in its production, we are not able to guarantee that the book is completely error-free. Additionally, while the book has been written to closely match the CCEA specification, it is the responsibility of each candidate to satisfy themselves that they have fully met the requirements of the CCEA specification prior to sitting an exam set by that body. For this reason, and because specifications change with time, we strongly advise every candidate to avail of a qualified teacher and to check the contents of the most recent specification for themselves prior to the exam. Colourpoint Creative Ltd therefore cannot be held responsible for any errors or omissions in this book or any consequences thereof.

Maths Revision Booklet
for CCEA GCSE 2-tier specification

M3

Neill Hamilton

Contents

A calculator may be used in these exercises.

Revision Exercise 1 3

Revision Exercise 2 12

Revision Exercise 3 23

Revision Exercise 4 31

Answers 42

Dedicated to the 5 best grandchildren in the world Cadence, Lily, Willow, Quinn and Rory and the best marathon runner ever, David.

Revision Exercise 1

1. The exam marks of 8 pupils in History and English are shown below:

History	27	42	56	38	64	72	48	16
English	51	62	86	59	91	92	74	39

   (a) Draw a scatter graph of this data on the grid below.

   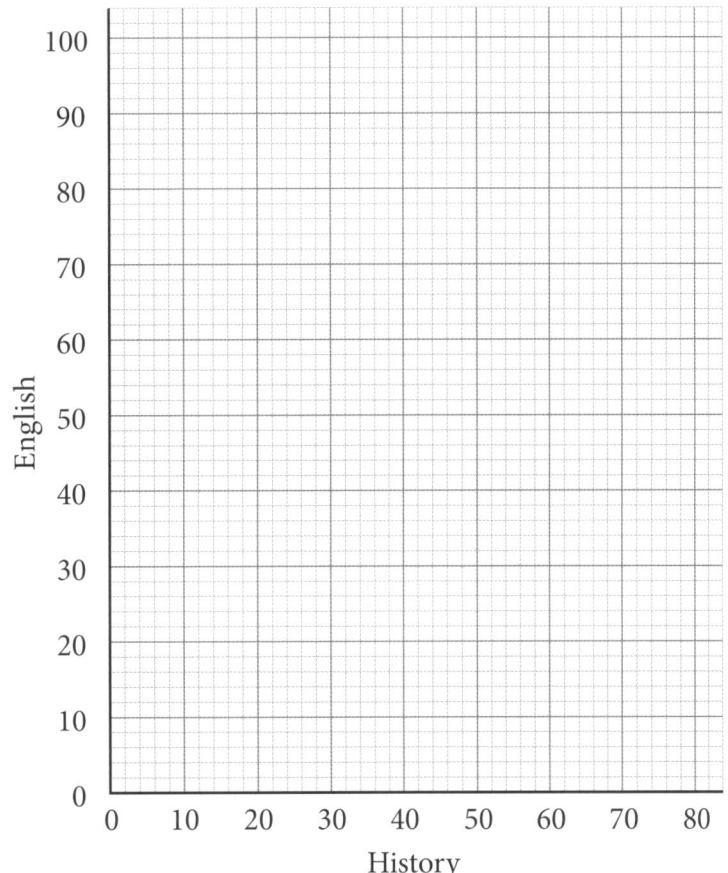

   (b) Draw a line of best fit on the grid.

   (c) Describe the correlation.

   Answer _____

   (d) A pupil scored 34 in History but she did not do her English exam. Use the line of best fit to estimate the mark she might have scored in English, had she sat the exam.

   Answer _____

2. The area of the triangle below is 54 cm². Find x.

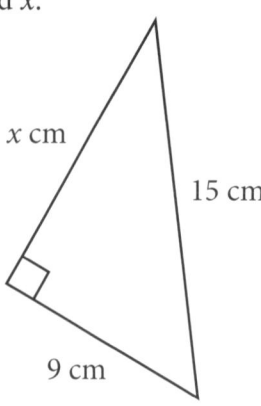

Answer $x =$ _____ cm

3. Ann scored 111 out of 150 in Science. What was her percentage score?

Answer _____ %

4. The average attendance at a football match last year was 1600.
This year the average attendance rose by 15%.
Work out the average attendance this year.

Answer _____

5. (a) In 2018 there was an 8% increase in the number of pupils attending gymnastics courses in the UK.
If 8100 attended the courses in 2018, how many attended in 2017?

Answer _____

(b) In 2019 there was a 8% decrease in the number of pupils attending the courses. Explain why the numbers attending the courses in 2017 and 2018 are not the same.

Answer _____

_____

_____

Revision Exercise 1

6. In the diagram below, the lines AB and CD are parallel. Work out the angles *x* and *y*.

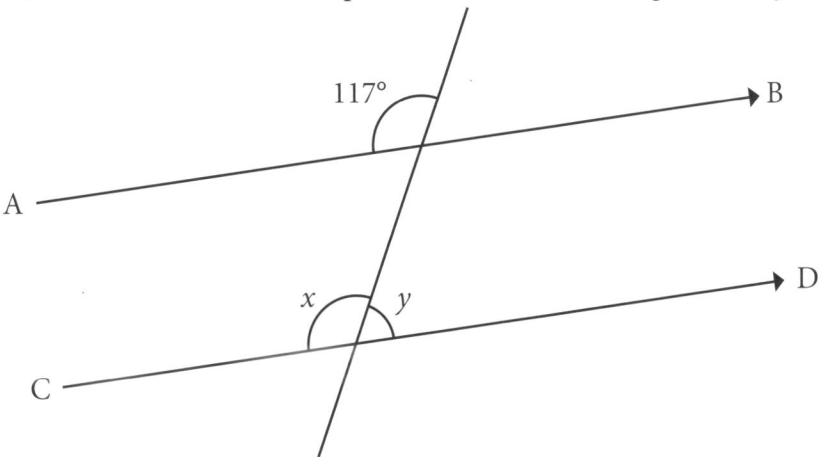

Answer *x* = _____ ° and *y* = _____ °

7. Colin bought *x* pens costing 70p each and *y* pencils costing 36p each.
   (a) Write down an expression for the total cost of the pens and pencils. Give your answer in pence.
   (Hint: Remember to put the numbers first in algebra.)

   Answer _____ p

   (b) Colin paid with a £20 note. Write down an expression for his change. Give your answer in £.

   Answer £ _____

8. Solve the following equations:
   (a) $5x - 4 = 2x + 8$

   Answer *x* = _____

   (b) $4(x - 5) = 24$

   Answer *x* = _____

   (c) $4(2x - 3) = 5(x - 4)$

   Answer *x* = _____

9. A teacher asked a class to estimate when they thought one minute had passed.
   The results are given below in seconds.

   54 57 63 73 58 79 61 50 66 74 49 57 69 63 54 68 72 57 52 45

   (a) Draw a stem and leaf diagram to show this data.
       (Hint: The numbers must be put in order. You may need to make two drawings. Do not forget the key.)

   (b) Use your diagram to find:
       (i) the range of the estimates;

       Answer _____ s

       (ii) the median of the estimates.

       Answer _____ s

10. A package consists of b boxes each weighing 250 grams and 7 cartons each weighing 120 grams.
    The total weight of the boxes and cartons is 3.09 kg. Form an equation in b and solve it to find the value of b.
    (Hint: You must use algebra to do this question.)

    Answer $b =$ _____

11. Multiply out and simplify:
    (a) $3(2x - 5) + 4(x + 5)$

    Answer _____

    (b) $2(7 - 3x) - 3(2x - 5)$

    Answer _____

Revision Exercise 1

12. (a) Find the coordinates of the midpoint of the line joining PQ in the grid below.

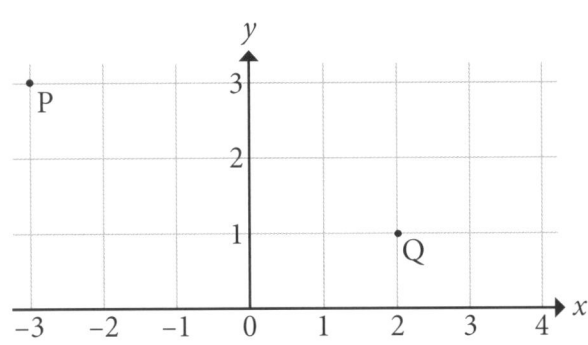

Answer (_____ , _____)

(b) The length of the line PQ.

Answer _____

13. Factorise completely:
 (a) $4x + 10$

Answer _____

 (b) $vy - v^2$

Answer _____

 (c) $10t^3 - 15t^2h$

Answer _____

14. A width was measured as 9.7 cm, correct to the nearest millimetre. Write down the least value this width could be.

Answer _____ cm

15. The formula to change °F to °C is: $C = \frac{5}{9}(F - 32)$
 (a) Use this formula to change:
  (i) 77°F to °C

Answer _____ °C

  (ii) 86°C to °F

Answer _____ °F

(iii) 14°F to °C

Answer _____ °C

(b) Find the value of C for which F = C.

Answer _____

16. The equation of a straight line is $y = 5x - 3$
    Find where this line crosses:
    (a) the y-axis

    Answer (_____ , _____)

    (b) the x-axis

    Answer (_____ , _____)

    (c) the line $y = 7$

    Answer (_____ , _____)

17. Expand:
    (a) $x(x^2 + 3)$

    Answer _____

    (b) $4y(y^2 - 3y)$

    Answer _____

    (c) $n^2(n + m)$

    Answer _____

18. A, B and C are three points on horizontal ground, as shown below.

    A —— 100 m —— B —— 150 m —— C

    There is a vertical tower with its base at B.
    The angle of elevation of the top of the tower from A is 26°
    Work out the angle of elevation of the top of the tower from C.

    Answer _____

Revision Exercise 1

19. (a) Each pupil in a class chooses one place to go on a school trip from the following list:

Portrush   *Ulster Museum Belfast*   *Tollymore Forest Park*   *Armagh Planetarium*

  (i) Which would be the most appropriate average for the teacher to use to determine where the class should go for the school trip?

Answer _____

  (ii) Explain why you chose this average.

Answer _____

(b) A teacher marks all the pupils in her class in a Mental Maths test.

  (i) Which would be the most appropriate average for her to use when writing a school report?

Answer _____

  (ii) Explain why you chose this average.

Answer _____

20. (a) Write the following as products of prime factors:
  (i) 60

Answer _____

  (ii) 90

Answer _____

(b) Hence find:
  (i) the HCF of 60 and 90

Answer _____

  (ii) the LCM of 60 and 90

Answer _____

21. Expand and simplify:
    (a) $5(2x - 3) + 3(4 - x)$

    Answer _____

    (b) $3(4x - 3) - 2(x - 7)$

    Answer _____

22. Solve the following:

    (a) $\dfrac{k}{3} - 6 = 4$

    Answer $k = $ _____

    (b) $\dfrac{5v + 3}{4} = 2v - 3$

    Answer $v = $ _____

    (c) $\dfrac{5 - 2x}{3} = \dfrac{3x - 5}{2}$

    Answer $x = $ _____

23. A shopkeeper bought 42 boxes of apples for £188.
    Each box held 25 apples.
    She sold ⁷⁄₁₀ of the apples at 24p each.
    She sold 12% of the apples at 18p each.
    She sold the rest of the apples at 14p each.

    (a) Work out her total profit.

    Answer _____

    (b) Work out her total profit as a percentage of the cost price. Give your answer to 2 decimal places.

    Answer _____

24. Thomas saved £P from his pay last week. Vicky saved three times as much as Thomas. William saved £25 less than Vicky. Altogether they saved £171. Form an equation and solve it to find P.
(Hint: You must use algebra to do this question.)

Answer P = _____

25. **(a)** Complete the table below for y + 2x = 8.

x	−2	0	2	4
y				

(b) Draw the straight line y + 2x = 8 on the grid below.

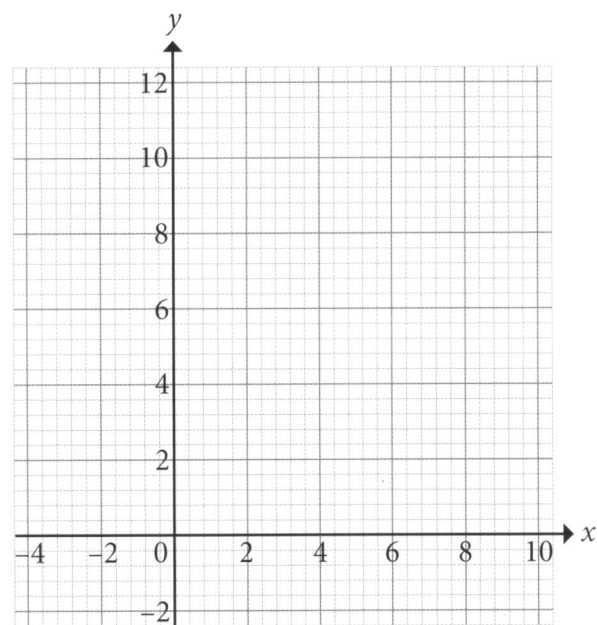

26. PQ is a line where P = (6, −2) and Q = (2, −5). Find:
(a) M, the midpoint of PQ

Answer (_____ , _____)

(b) the length of PQ

Answer _____

Revision Exercise 2

1. **(a)** Draw the graph of $y = 5x - 2$ on the grid below.

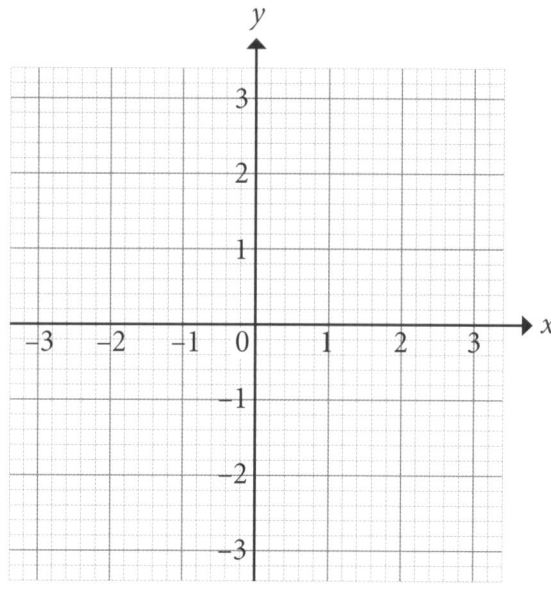

(b) Explain the difference between the graph $y = 5x - 2$ and the graph of:
   (i) $y = 5x$
   (ii) $y = -5x - 2$
   without drawing these graphs.

Answer **(i)** _____

_____

Answer **(ii)** _____

_____

2. There are 20 pupils in a class.
   The ratio of boys to girls in the class is 3:2.
   The mean height per pupil in the class is 1.572 m.
   The mean height per girl in the class is 1.47 m.
   Work out the mean height per boy in the class.

Answer _____

Revision Exercise 2

3. The table below shows the marks of 300 pupils in an exam.

Mark (m)	$0 < m \leq 10$	$10 < m \leq 20$	$20 < m \leq 30$	$30 < m \leq 40$	$40 < m \leq 50$	$50 < m \leq 60$	$60 < m \leq 70$
Frequency	15	35	55	85	50	40	20

(a) Complete the cumulative table below.

Mark (less than or equal to)	0	10	20	30	40	50	60	70
Cumulative Frequency	0	15	50					300

(b) Draw the cumulative frequency curve on the grid below.

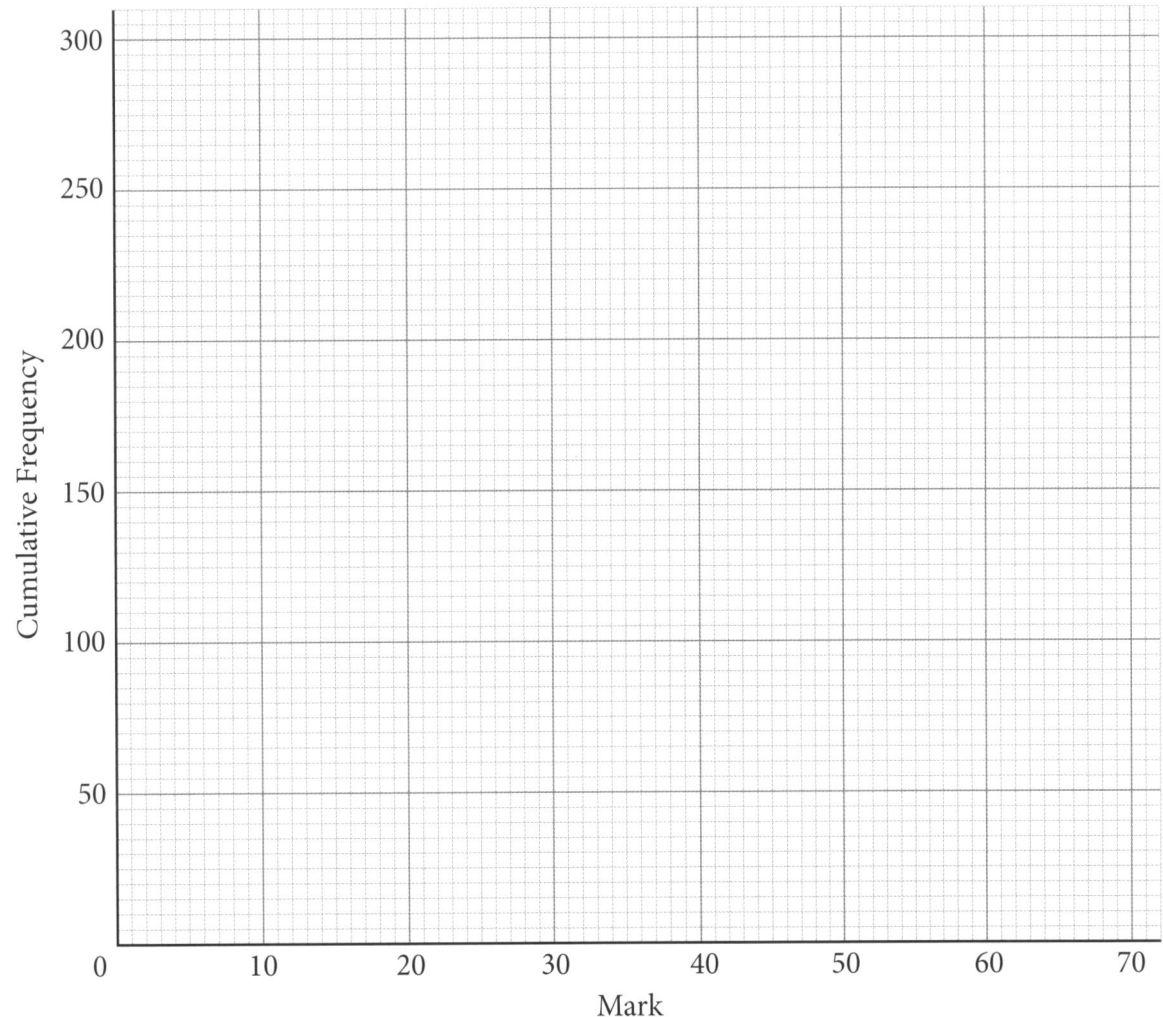

(c) Use your cumulative frequency curve to estimate the:
   (i) median

   Answer _____

   (ii) inter-quartile range

   Answer _____

   (iii) pass mark if 45% of the pupils pass the exam.

   Answer _____

4. The marks of a class in Science and Geography were recorded.
   The table below shows some facts about the Geography marks.

Minimum Mark	14
Lower Quartile	22
Median	25
Upper Quartile	28
Maximum	34

   (a) The box plot below shows the Science marks.
       Draw a box plot to show the Geography marks on the same grid.

   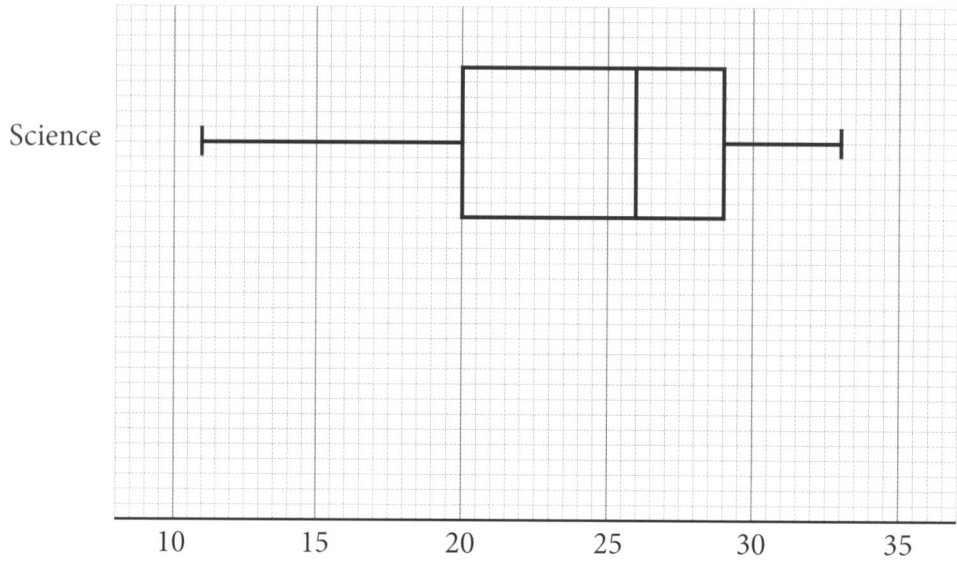

   (b) Write two statements comparing the two sets of marks.

   _____

   _____

   _____

   _____

   _____

   _____

Revision Exercise 2

5. Factorise completely:
   (a) $10x^2y^3 - 15xy$

   Answer _____

   (b) $t^2 - 9t + 20$

   Answer _____

   (c) $m^2 - 64$

   Answer _____

   (d) $k^2 + k - 12$

   Answer _____

6. Cadence thinks of a number.
   She trebles it and then takes away 7 from her answer.
   She then doubles this answer and gets 13
   Show, by algebra, that the number she thought of was 4.5

   Answer:

7. (a) Simplify $\dfrac{8x^3y^2}{4xy}$

   Answer _____

   (b) Simplify $\dfrac{x}{y} + \dfrac{y}{z}$

   Answer _____

8. (a) Write the following numbers as products of prime factors:
   (i) 90

   Answer _____

   (ii) 525

   Answer _____

   (b) Find:
   (i) the HCF of 90 and 525

   Answer _____

   (ii) the LCM of 90 and 525

   Answer _____

   (c) What is the smallest integer n (bigger than 1) for which $525n$ is a square number?

   Answer _____

9. In the diagram below, AB and CD are parallel lines.

   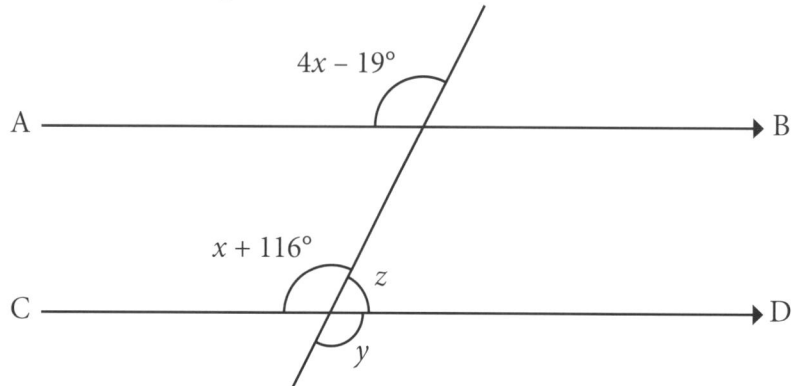

   (a) Explain why $4x - 19 = x + 116$.

   Answer _____

   _____

   (b) Hence find the value of:
   (i) x

   Answer $x =$ _____

   (ii) y

   Answer $y =$ _____

   (iii) z

   Answer $z =$ _____

Revision Exercise 2

10. Solve the following:

(a) $\frac{1}{2}(x+1) - \frac{1}{3}(x-1) = 2$

Answer $x =$ _____

(b) $\frac{4x-1}{3} + \frac{2-x}{2} = 4$

Answer $x =$ _____

(c) $\frac{15-2x}{3} = 12 - 3x$

Answer $x =$ _____

11. Simplify:

(a) $\frac{4}{f} + \frac{f}{4}$

Answer _____

(b) $\frac{p}{q} \times \frac{r}{q}$

Answer _____

(c) $\frac{5}{h} \div \frac{15}{h}$

Answer _____

12. Martin earns £975.60 per month. He pays Income Tax as follows:

The first £5700 of his annual salary is tax free.
He pays 22% Income Tax on the next £3850 of his annual salary.
He pays 40% Income Tax on the remainder of his annual salary.

Calculate how much tax Martin pays in a year.

Answer £ _____

13. Arlene deposits £x in a bank which pays y% compound interest per year.
    After one year her total amount including interest is £82
    After two years her total amount including interest is £84.05
    Find:
    (a) y

    Answer y = _____

    (b) x

    Answer x = _____

14. Explain how a crayon, measured at 11 cm long, correct to the nearest centimetre, might be able to be placed inside a rectangular box measuring 4 cm by 9 cm, each measured correct to the nearest centimetre.

    _____
    _____
    _____
    _____
    _____

15. A boat sails 11.4 km due north of harbour H and then 15.2 km due west until it comes to harbour P.
    Another boat sails 4 km due east of harbour H and then 9.6 km due north until it comes to harbour Q.
    Work out the distance PQ.

    Answer _____

16. There are 100 people in a youth club.
   x played tennis, netball and badminton.
   11 played tennis and netball.
   9 played tennis and badminton.
   12 played netball and badminton.
   22 played only tennis.
   24 played only netball.
   20 played only badminton.
   14 played none of these three sports.

   (a) Show this information on a Venn diagram.

   (b) Hence find the value of x, i.e. how many of these people played all three of these sports.

   Answer _____

   (c) How many of these people played Badminton?

   Answer _____

17. Bill bought a bike for £164. He sold it a year later at a loss of 24%.
Calculate his selling price.

Answer £ _____

18. A circle has diameter 8.6 cm. Calculate:
(a) its circumference
(b) its area.
(**Remember to include units in your answers.**)

Answer (a) Circumference = _____ (b) Area = _____

19. An electricity bill is shown below.

Units Used	Cost Per Unit (pence)	Fixed Charge (£)	Total Cost (£)
5478	2.8	31.56	

(a) Work out the total cost in £.
(**Hint: Be careful with the units. Check that the answer looks sensible. Round your answers appropriately.**)

Answer £ _____

(b) VAT at 5% is then added to the total cost to determine the price charged for electricity.
Work out the total price charged.

Answer £ _____

20. The point A has coordinates (−2, 1) and the point B has coordinates (4, −7).
Line L bisects AB and is parallel to the line with equation $2y - 4x = 7$
Find where the line L crosses the x-axis.

Answer _____

Revision Exercise 2

21. I can buy a shirt from web site A for £18.40 but also have to pay VAT at 20%.
Alternatively, I can pay £20.97 but no VAT from web site B.
Which is cheaper and by how much?
Show your working.

Answer _____

22. The lengths, l cm, of 80 objects are given below.

Length (l)	$0 < l \leq 10$	$10 < l \leq 20$	$20 < l \leq 30$	$30 < l \leq 40$	$40 < l \leq 50$
Frequency	15	17	21	19	8

Work out an estimate for the mean length.

Answer _____ cm

23. Cormac earns £340 per week.
(a) What does he earn in a year?

Answer £ _____

(b) He spends 15% of his weekly pay on food. How much does he spend on food weekly?

Answer £ _____

(c) He saves £13.50 per week.
  (i) How many weeks will it take him to save up to buy a television priced at £380?

Answer _____ weeks

  (ii) How much of his savings will he have left after buying the television?

Answer £ _____

24. (a) Work out the area shaded between the square and circle shown in the diagram below left.
    (b) Work out the total area shaded in the rectangle and quarter circle shown in the diagram below right.

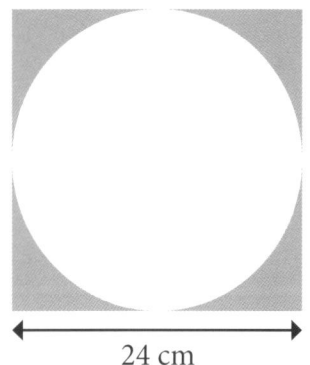
24 cm

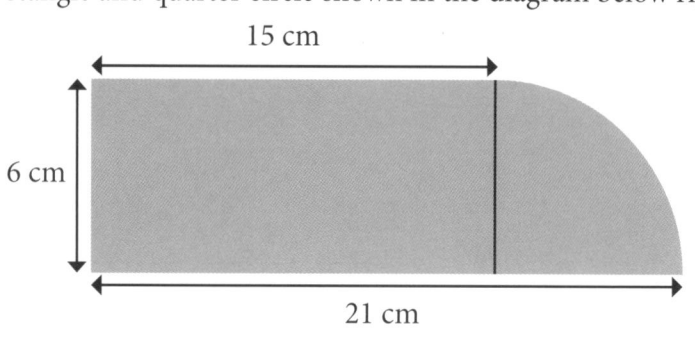
15 cm
6 cm
21 cm

Answer (a) _____ cm²

Answer (b) _____ cm²

25. A semicircle has diameter 12 cm. Calculate:
(a) its perimeter

Answer _____ cm

(b) its area.

Answer _____ cm²

24. ABCD is a rectangle. AB = 18.2 cm and AC = 65 cm. Find the length of BC.
(**Hint: Name each vertex of the rectangle in turn in the order given, i.e. A then B then C then D.**)

Answer _____ cm

25. The price of petrol rose from £1.239 per litre to £1.485 per litre over three months. Find the percentage rise correct to one decimal place.

Answer _____ %

Revision Exercise 3

1. A ship starts at H and sails 23.2 km due north and then 17.4 km due west.
   How far is the ship from its starting point H? **(Hint: Draw a sketch.)**

   Answer _____ km

2. A cylinder has base diameter 16 cm and height 7.5 cm.
   Calculate its volume in litres.
   Give your answer correct to two decimal places.
   (Hint: You have to learn this formula. It is not given on the formula sheet.)

   Answer _____ litres

3. Calculate what £600 will amount to in 3 years at 2.5% compound interest per year.
   (Hint: Give your answer to the nearest penny.)

   Answer £ _____

4. Rory's electricity tariff is shown below.

COST OF ELECTRICITY
Fixed charge £15.79
Cost per unit 6.3p

   The cost of Rory's electricity was £52.33.
   How many units did he use?

   Answer _____

5. Calculate the volume of the isosceles triangular prism ABCDEF shown below.

   AB = AC = 24 cm   BC = 28.8 cm   CD = 4.6 m

   (Hint: You need to find the perpendicular height of triangle ABC. Make the units the same. The formula is given on the formula sheet.)

   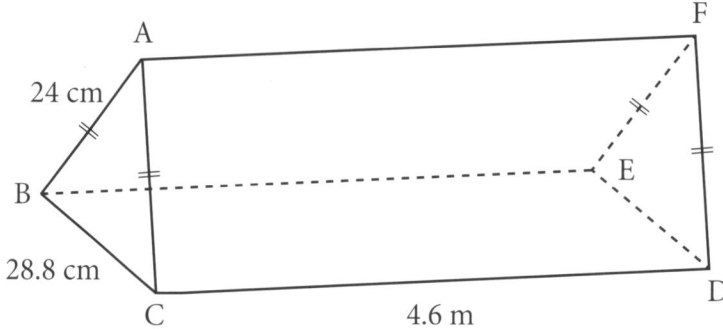

   Answer _____

6. A car depreciated by 24% in the first year, 20% in the second year and 18% in the third year. It cost £8600 to buy new. Find:
   (a) its value after 3 years

   Answer £ _____

   (b) the total depreciation as a percentage of the original price, to 1 decimal place.

   Answer _____ %

7. A prism of base area 176.8 cm² and height 5.7 cm is filled with water at a rate of 6.8 cm³ per second. How long does it take to fill the prism? (Hint: The formula for the volume is on the formula sheet.)

   Answer _____ s

Revision Exercise 3

8. (a) The VAT (at 20%) on a television was £92.20
   Find the price of this television before VAT is added.

   Answer £ _____

   (b) The selling price of a washing machine (including 20% VAT) was £388.80
   Find the price before VAT was added.

   Answer £ _____

9. (a) Work out the length of PQ in the diagram below.
   (Hint: Give your answer correct to 3 significant figures. Make sure your calculator is in degree mode.)

   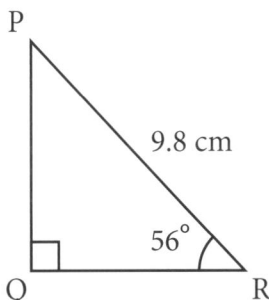

   Answer _____ cm

   (b) Find ∠YXZ in the diagram below.

   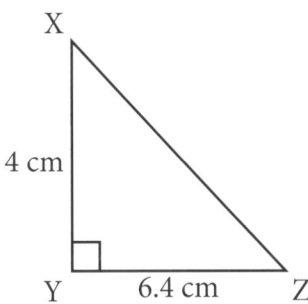

   Answer _____ °

10. The volume of a cuboid is 91.2 cm³. Its length is 4.8 cm and its breadth is 2.5 cm.
    Find its height.

    Answer _____ cm

11. The height and width of a box was measured as 6.8 cm and 4.7 cm, each correct to the nearest millimetre. Calculate the:
(a) greatest possible total height of 9 of these boxes

Answer _____ cm

(b) greatest possible total width of 7 of these boxes.

Answer _____ cm

12. Show that $(v + 2)^2 - (v - 1)^2 = 3(2v + 1)$

13. From a point A on the ground, the angle of elevation of the top of a vertical statue PQ is 25° as shown below. AQ = 8.4 m.

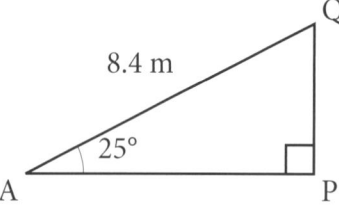

Calculate:
(a) PQ

Answer _____ m

(b) AP

Answer _____ m

14. The length and breadth of a rectangle were measured as 9.7 cm and 4.2 cm each correct to 1 decimal place. Calculate:
(a) the minimum possible perimeter

Answer _____ cm

(b) the maximum possible area.

Answer _____ cm²

Revision Exercise 3

15. The rectangle and square below both have the same perimeter.

    Rectangle: $6x - 1$ by $2x + 1$
    Square: side $5x - 3$

    (a) Form an equation in x and hence find the value of x.

    Answer $x =$ _____

    (b) Find the area of the rectangle as a percentage of the area of the square.
        Give your answer to one decimal place.

    Answer _____ %

16. The table below shows the numbers of goals scored in 60 football matches.

Number of goals	0	1	2	3	4	5
Frequency	5	9	18	22	4	2

Calculate the:
(a) modal number of goals scored,

Answer _____

(b) median number of goals scored,

Answer _____

(c) mean number of goals scored, correct to two decimal places.

Answer _____

17. Look at the flow diagram below.

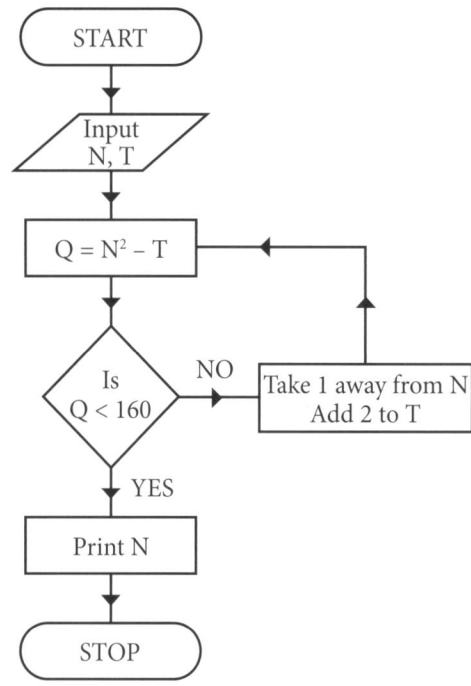

Start with N = 16 and T = 12
List the printed value on the answer line.
(Hint: You must show your work.)

Answer N = _____

18. M(−3, 4) is the midpoint of AB where A is the point (1, 2).
Find the coordinates of B.

Answer _____

19. The airport bus leaves the city hall every half hour.
The hospital bus leaves the city hall every 36 minutes.
They both leave together at 0720.
When will they next both leave together?

Answer _____

Revision Exercise 3

20. A rectangle has length $(x + 3)$ cm and breadth $(x - 2)$ cm.
The area of the rectangle is 12 cm² less than the area of a square side 6 cm.
(a) Form a quadratic equation in x.

Answer _____

(b) Solve this equation to find the value of x.

Answer _____

21. $(x + p)^2 + q \equiv x^2 - 6x + 3$

Work out the value of p and q.

Answer _____

22. Factorise:
(a) $x^2 - 81$

Answer _____

(b) $49 - x^2$

Answer _____

(c) $4x^2 - 25$

Answer _____

23. 12 pupils sat a literacy test. The top 11 marks in ascending order were:

7   8   9   11   12   13   14   14   17   20   24

The range of all 12 marks is 3 times the inter-quartile range of all 12 marks.
Find the lowest of all 12 marks.

Answer _____

24. Simplify:

(a) $\dfrac{10x^3y^3}{15x^2y^4}$

Answer _____

(b) $\dfrac{4x^3}{3y} \times \dfrac{6y^2}{x^2}$

Answer _____

(c) $\dfrac{p}{q} \times \dfrac{q}{r}$

Answer _____

(d) $\dfrac{x^2}{y^2} \div \dfrac{x^3}{y^3}$

Answer _____

25. Buses leave Lurgan for Newry every 1 hour 15 minutes.
Buses leave Lurgan for Armagh every 1 hour 40 minutes.
A bus leaves Lurgan for Newry at 9:20 am and another bus leaves Lurgan for Armagh at 9:20 am.
What is the next time that two buses will leave Lurgan at the same time for Newry and Armagh?

Answer _____

26. AOB is a sector of a circle centre O in which angle AOB is 35°.
The perimeter of the sector AOB is 20.89 cm.
Work out the area of the sector AOB to 2 decimal places.

Answer _____

27. Simplify:

(a) $\dfrac{3x-2}{6} + \dfrac{2x+3}{9}$

Answer _____

(b) $\dfrac{4x-3}{2} - \dfrac{2-3x}{4}$

Answer _____

Revision Exercise 4

1. The area of the rectangle below is 50 cm².

   (a) Form and solve a quadratic equation in x.

   Answer _____

   (b) Hence find the perimeter of this rectangle.

   Answer _____

2. Find the equation of the line L which passes through $(-1, 2)$ and $(3, -6)$.

   Answer _____

3. ABCD is a rectangle with AB = 14 cm and angle CAB = 24°.
   PQRS is a square.
   ABCD and PQRS have the same area.
   Work out the perimeter of PQRS to 1 decimal place.

   Answer _____

4. 50 people were surveyed about what they had for breakfast.
   3 had toast, porridge and yogurt.
   11 had toast and porridge.
   7 had toast and yogurt.
   8 had yogurt and porridge.
   25 had toast.
   23 had porridge.
   21 had yogurt.

   (a) Show this information on a Venn diagram.

   (b) Hence find how many of these people had none of the three items for breakfast.

   Answer _____

5. Find the equation of the line L which passes through $(2, -4)$ and which is parallel to the line $y = 6x - 3$.

   Answer _____

Revision Exercise 4

6. Which two of these lines are parallel?

   A $y = 3x - 4$
   B $2y + 6x = 3$
   C $3y = 6x + 1$
   D $y = 4 - 3x$

   Answer _____

7. (a) Draw a tangent to the circle below.

   (b) Show, by shading, a segment of the circle below.

   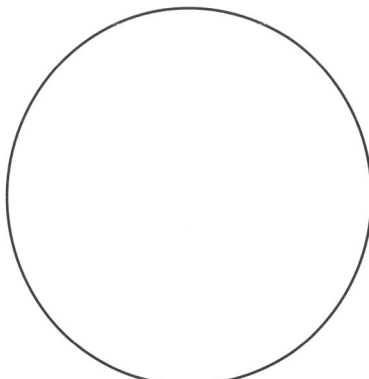

8. Work out the pressure exerted by a force of 310.08 N which acts on a rectangular metal plate of length 7.6 m and breadth 4.8 m.

   Answer _____

9. A force of 36 N exerts a pressure of 22 N/m² acting on a circular metal plate. Find the radius of the circular plate.

   Answer _____

10. O is the centre of a circle radius 6 cm, as shown below.

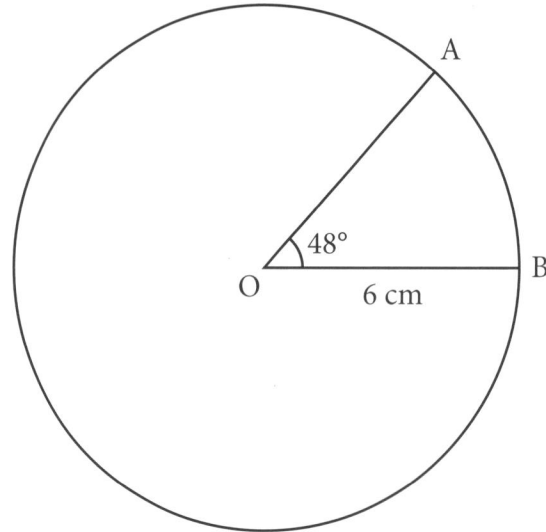

Work out:
(a) the length of the arc AB,

Answer _____

(b) the perimeter of the sector OAB,

Answer _____

(c) the area of the sector OAB.

Answer _____

11. The volume of a cylinder is 60 cm³.
The curved surface area of the cylinder is 75 cm².
Work out the radius of the cylinder.

Answer _____

12. O is the centre of a circle. The arc AB is 9.425 cm long.

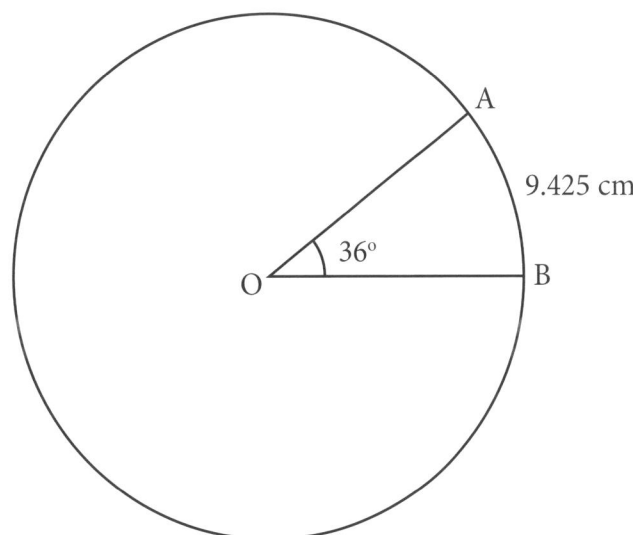

Work out:
(a) the radius of the circle to the nearest integer,

Answer _____

(b) the area of the sector OAB to 2 decimal places.

Answer _____

13. The table below shows the lengths of different objects, measured to the nearest centimetre.

Length (cm)	Frequency
1–3	7
4–6	4
7–9	x
10–12	3

The mean length is 5.75 cm.
Work out the value of x.

Answer _____

14. The box plot below shows the marks of a class in a RE test.

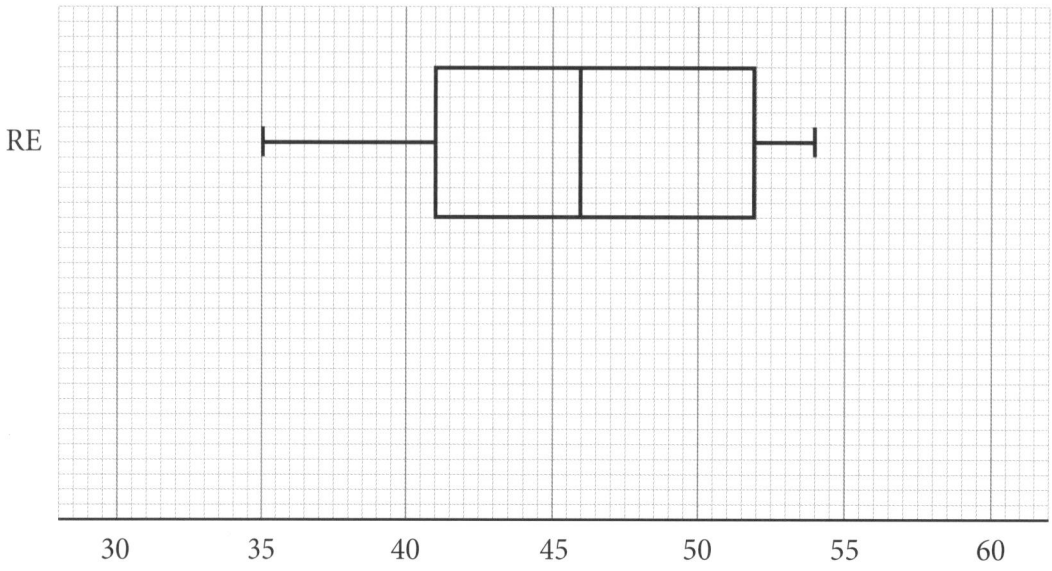

The marks for the same class in an LLW test were as follows:

32	45
35	47
36	47
36	49
37	51
38	52
41	53
43	57
44	57

(a) Draw the box plot for the marks in LLW on the grid above.

(b) Hence make two different comparisons about the marks in RE and LLW.

Answer

Revision Exercise 4

15. A cylinder open at one end has base radius 4.2 cm and total surface area 960 cm².
    Work out:
    (a) its curved surface area,

    Answer _____

    (b) its height,

    Answer _____

    (c) its volume.

    Answer _____

16. A hemisphere has diameter 18 cm.
    Work out:
    (a) its total surface area,

    Answer _____

    (b) its volume.

    Answer _____

17. The volume of sphere A is 8 times the volume of sphere B.
    How many times bigger is the surface area of A than B?

    Answer _____

18. A metal cylinder of radius 1.6 m and height 2.4 m is to be melted down and recast into identical metal spheres each of radius 35 cm.
    How many spheres can be made?

    Answer _____

19. A hollow cone has perpendicular height 12 cm and base radius 10 cm, as shown below.

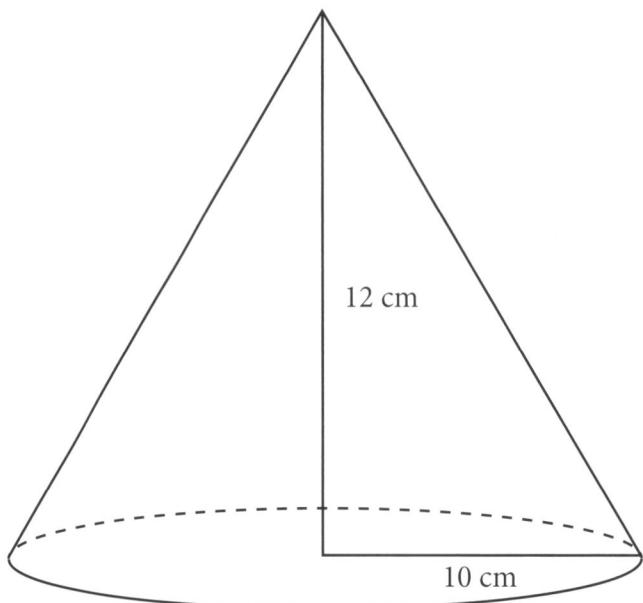

Work out:
(a) its volume,

Answer _____ cm

(b) its total surface area.

Answer _____ cm²

20. The total surface area of a solid cone, of base radius 6 cm, is 720 cm². Work out:
(a) the curved surface area of the cone,

Answer _____

(b) the perpendicular height of the cone to 1 decimal place,

Answer _____

(c) the volume of the cone.

Answer _____

Revision Exercise 4

21. Cone A has base radius a cm and perpendicular height b cm.
    Cone B has base radius $3a$ cm.
    The volume of cone A is double the volume of cone B.
    Work out the perpendicular height of cone B in terms of b.

    Answer _____

22. The marks of 20 pupils in a test are as follows:

    2   6   7   9   10   12   13   14   14   16   17   19   20   21   22   26   27   27   28   30

    Find:
    (a) the median,

    Answer _____

    (b) the lower quartile,

    Answer _____

    (c) the upper quartile,

    Answer _____

    (d) the interquartile range.

    Answer _____

23. Write down six different numbers which together have a median of 10, an interquartile range of 7 and a range of 15.

    Answer _____

24. TQ is a vertical tower with T on the ground, as shown below.
A is on the ground and 24 m west of T.
B is on the ground and 40 m east of T.
The angle of elevation of Q from A is 68°.

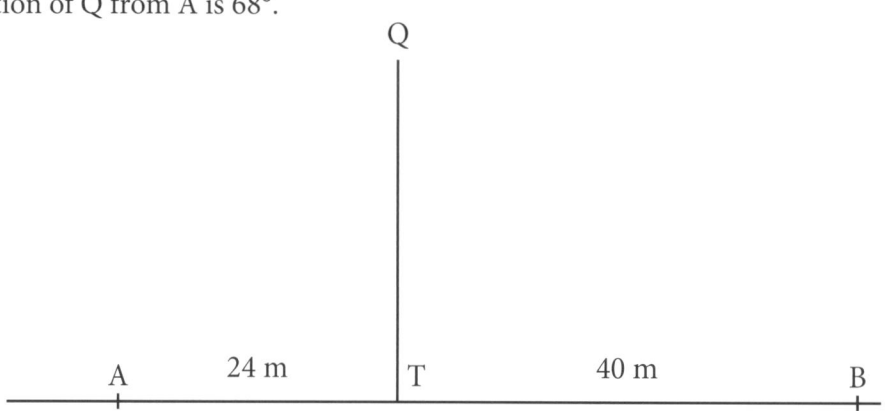

Find the angle of elevation of Q from B.

Answer _____

25. A metal is in the shape of a trapezium as shown below.

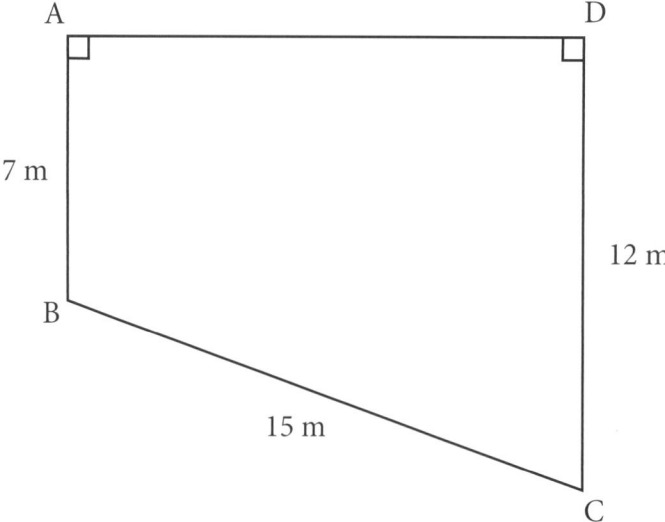

(a) Work out the area of the trapezium.

Answer _____

(b) Hence work out the force needed to be applied to this metal to exert a pressure of 2000 N/m².

Answer _____

Revision Exercise 4

26. ABCD is a rectangular field with AB = 18 m.
A hedge runs from A to C as shown.
The shortest distance from D to the hedge is 4.5 m.
Work out the area of the field.

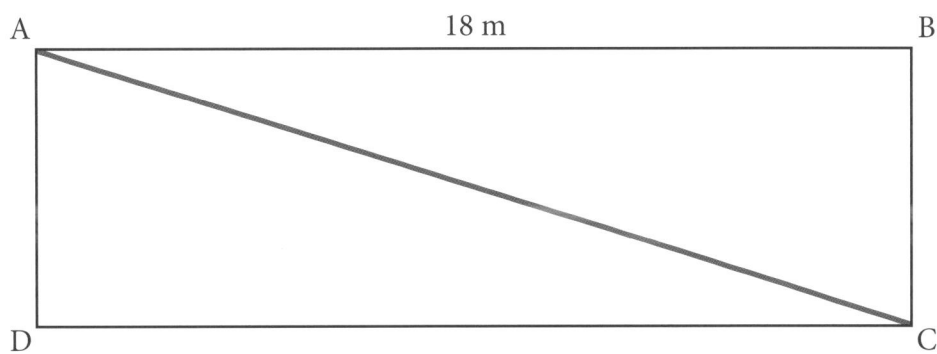

Answer _____

27. The base radius and height of a cylinder were measured as 6 m and 14 m each correct to the nearest integer.
Work out:
(a) the minimum possible curved surface area,

Answer _____

(b) the maximum possible volume.

Answer _____

Answers

Revision Exercise 1

1. (c) Positive (d) Approx 58 (depending on line of best fit drawn on graph)
2. 12 (by Pythagoras' theorem)
3. 74%
4. $1600 + (1600 \times 15 \div 100) = 1840$
5. (a) $8100 \div 1.08 = 7500$ (b) 8% of $8100 = 648$ but $8100 - 648 = 7452$ which doesn't equal 7500
6. $x = 117°$, $y = 180 - 117 = 63°$
7. (a) $70x + 36y$ (b) £$20 - 0.7x - 0.36y$
8. (a) 4 (b) 11 (c) $-8/3$
9. (a)

4	5	9					
5	0	2	4	4	7	7	8
6	1	3	3	6	8	9	
7	2	3	4	9			

   Key 4 | 5 = 45 seconds
   (b) (i) $79 - 45 = 34$ seconds
   (ii) $(58 + 61) \div 2 = 59.5$ seconds
10. $0.25b + (7 \times 0.12) = 3.09$; giving $b = 9$
11. (a) $10x + 5$ (b) $29 - 12x$
12. (a) $(-0.5, 2)$ (b) 5.385 (by Pythagoras' theorem)
13. (a) $2(2x + 5)$ (b) $v(y - v)$ (c) $5t^2(2t - 3h)$
14. $9.7 - 0.05 = 9.65$
15. (a) (i) $C = 5/9(77 - 32) = 25°C$ (ii) $86 = 5/9(F - 32)$; giving $F = 186.8°F$ (iii) $C = 5/9(14 - 32) = -10°C$
    (b) If $F = C$ then $C = 5/9(C - 32)$; giving $C = -40$
16. (a) $x = 0$ gives $y = -3$; $(0, -3)$ (b) $y = 0$ gives $x = 0.6$; $(0.6, 0)$ (c) $y = 7$ gives $x = 2$; $(2, 7)$
17. (a) $x^3 + 3x$ (b) $4y^3 - 12y^2$ (c) $n^3 + n^2 m$
18. Find height: $\tan 26 = BD \div 100$; so $BD = 48.77$ m; Then: $\tan C = 48.77 \div 150$; giving $C = 18°$
19. (a) (i) Mode (ii) The trip should be to the most popular place. (b) (i) Median (ii) This shows whether the pupil is in the top or bottom half of the class.
20. (a) (i) $2 \times 2 \times 3 \times 5$ (ii) $2 \times 3 \times 3 \times 5$
    (b) (i) $2 \times 3 \times 5 = 30$ (ii) $2 \times 2 \times 3 \times 3 \times 5 = 180$
21. (a) $7x - 3$ (b) $10x + 5$
22. (a) 30 (b) × both sides by 4: $5v + 3 = 8v - 12$; giving $v = 5$ (c) × both sides by 6: $10 - 4x = 9x - 15$; giving $x = 25/13$
23. (a) Total apples $= 42 \times 25 = 1050$; $7/10 \times 1050 = 735$; 12% of $1050 = 126$; so there are 189 apples left; total income $= (735 \times 0.24) + (126 \times 0.18) + (189 \times 0.14) = £225.54$; $225.54 - 188 = £37.54$ (b) 19.97%
24. $P + 3P + (3P - 25) = 171$; giving $P = 28$
25. 12, 8, 4, 0
26. (a) $(4, -3.5)$ (b) 5 (by Pythagoras' theorem)

Revision Exercise 2

1. (a)

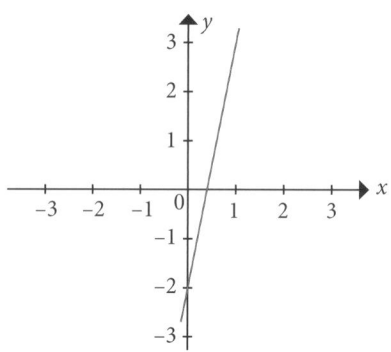

   (b)(i) The straight line goes through $(0, 0)$.
   (ii) The straight line slopes down from left to right.
2. Total number of girls $= 2/5$ of $20 = 8$; total of class heights $= 1.572 \times 20 = 31.44$; total of girl heights $= 1.47 \times 8 = 11.76$; if mean height of boy $= x$ then: $12x = 31.44 - 11.76$; giving $x = 1.64$ m
3. (a) 105 190 240 280 (b) Check answer from pupil's graph (c) (i) in the range 34.5 to 36.5 (ii) in the range 20.5 to 22.5 (iii) in the range 36 to 38
4. (a)

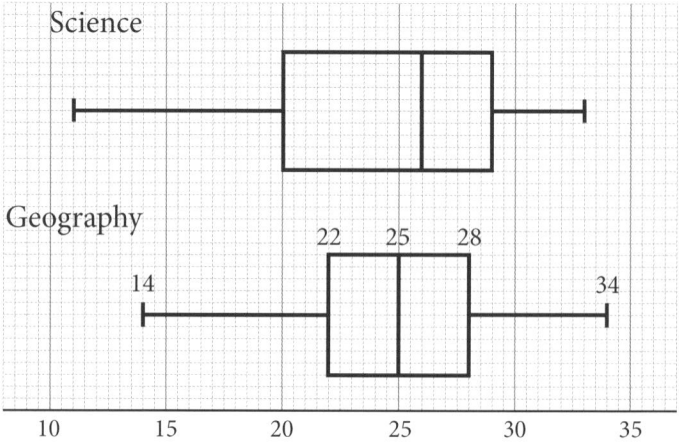

   (b) The range of Science marks is bigger than the range of Geography marks. The upper quartile in Science is bigger than the upper quartile in Geography. The lower quartile in Science is smaller than the lower quartile in Geography.

Answers

5. (a) $5xy(2xy^2 - 3)$ (b) $(t - 4)(t - 5)$
   (c) $(m - 8)(m + 8)$ (d) $(k + 4)(k - 3)$
6. $2(3x - 7) = 13$; $6x - 14 = 13$; $6x = 27$; giving $x = 4.5$
7. (a) $2x^2y$ (b) $\dfrac{xz + y^2}{yz}$
8. (a)(i) $2 \times 3 \times 3 \times 5$ (ii) $3 \times 5 \times 5 \times 7$ (b)(i) $3 \times 5 = 15$ (ii) $2 \times 3 \times 3 \times 5 \times 5 \times 7 = 3150$ (c) $525 = 3 \times 5^2 \times 7 = 5^2 \times 21$; so n cannot be less than 21
9. (a) Corresponding angles are equal
   (b) (i) 45 (ii) $45 + 116 = 161°$ (iii) $180 - 161 = 19°$
10. (a) $3x + 6 - 2x + 2 = 12$; giving $x = 7$
    (b) $8x - 2 + 6 - 3x = 24$; giving $x = 4$
    (c) $15 - 2x = 36 - 9x$; giving $x = 3$
11. (a) × both sides by $4f$, giving: $\dfrac{16 + f^2}{4f}$
    (b) × both sides by q, giving: $\dfrac{pr}{q^2}$ (c) × both sides by h, giving: ⅓
12. Annual wage after tax = $(975.6 \times 12) - 5700$
    = £6007.20; 22% tax on £3850 = £847; 6007 − 3850
    = £2157.20 left; 40% tax on £2157.20 = £862.88;
    total tax = 847 + 862.88 = £1709.88
13. (a) After 1 year: $82 = x \times (1 + y)$; after 2 years:
    $84.05 = x \times (1 + y)^2$; Dividing equation 2 by equation 1 gives: $y = 0.025 = 2.5\%$ (b) £80
14. Max dimensions are 4.5 cm and 9.5 cm
    $x^2 = 4.5^2 + 9.5^2$
    $x^2 = 110.5$
    $x = 10.51$
    Minimum crayon length = 10.5, and 10.5 < 10.51
15. East-west distance moved = $15.2 + 4 = 19.2$; north-south distance moved = $11.4 - 9.6 = 1.8$; so total distance PQ = 19.3 km by Pythagoras' theorem.
16. (a) 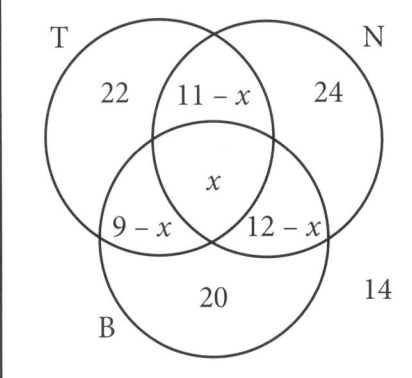 (b) 6 (c) 35
17. Loss = $164 \times 0.24 = 39.36$; $164 - 39.36 = £124.64$
18. (a) 27 cm (b) 58 cm²
19. (a) $(5478 \times 0.028) + 31.56 = £184.94$
    (b) $184.94 \times 1.05 = £194.19$
20. midpoint of AB = $(1, -3)$; $2y - 4x = 7$; so $y = 2x + 3.5$; so gradient of L = 2; using $y - y_1 = m(x - x_1)$ gives: $y - (-3) = 2(x - 1)$; so $y = 2x - 5$; then when $y = 0$, $x = 2½$; so point is $(2½, 0)$
21. Web site B by £1.11
22. Total of all lengths = $(15 \times 5) + (17 \times 15) + (21 \times 25) + (19 \times 35) + (8 \times 45) = 1880$; $1880 \div 80 = 23.5$
23. (a) £17 680 (b) £51 (c)(i) $380 \div 13.50 = 28.1$; i.e. 29 weeks (ii) $(29 \times 13.50) - 380 = £11.50$
24. (a) Square area − circle area = $24^2 - \pi \times 12^2$
    = $576 - 452.4 = 124$ to nearest cm (b) Quarter circle area = $¼\pi \times 6^2 = 28.3$; area rectangle = $15 \times 6 = 90$; total = $28.3 + 90 = 118$ to nearest cm
25. (a) $½\pi d + 12 = 30.8$ (b) $\pi r^2 \div 2 = 56.5$
26. 62.4 (by Pythagoras' theorem)
27. Difference = $1.485 - 1.239 = 0.246$ p;
    $0.246 \div 1.239 = 0.199 = 19.9\%$

Revision Exercise 3

1. 29 (by Pythagoras' theorem)
2. Base area × height = $\pi \times 8^2 \times 7.5 = 1.51$
3. $600 \times 1.025 \times 1.025 \times 1.025 = £646.13$
4. $15.79 + 0.063x = 52.33$; giving $x = 580$ units
5. Perpendicular height of ABC = 19.2 cm by Pythagoras' theorem; area of ABC
   = $½ \times 28.8 \times 19.2 = 276.48$; volume of prism
   = $276.48 \times 460 = 127180.8$ cm³ or 0.1271808 m³
6. (a) $8600 \times 0.76 \times 0.80 \times 0.82 = £4287.62$
   (b) Depreciation = $8600 - 4287.62 = 4312.34$;
   $4312.34 \div 8600 \times 100\% = 50.1\%$
7. Vol = $176.8 \times 5.7 = 1007.76$; $1007.76 \div 6.8 = 148.2$ s
8. (a) $92.20 \div 20 \times 100 = £461$ (b) $388.80 \div 1.20 = £324$
9. (a) $\sin 56 = PQ \div 9.8$; giving PQ = 8.12
   (b) $\tan X = 6.4 \div 4$; giving X = 58.0°
10. $91.2 \div 4.8 \div 2.5 = 7.6$
11. (a) $6.85 \times 9 = 61.65$ cm (b) $4.75 \times 7 = 33.25$ cm
12. $(v + 2)^2 = v^2 + 4v + 4$; and $(v - 1)^2 = v^2 - 2v + 1$
    So: $(v + 2)^2 - (v - 1)^2 = (v^2 + 4v + 4) - (v^2 - 2v + 1)$
    = $6v + 3 = 3(2v + 1)$
13. (a) $\sin 25 = PQ \div 8.4$; giving PQ = 3.55
    (b) 7.61 (by Pythagoras' theorem)
14. (a) $9.65 + 9.65 + 4.15 + 4.15 = 27.6$
    (b) $9.75 \times 4.25 = 41.4375$
15. (a) $2(2x + 1) + 2(6x - 1) = 4(5x - 3)$
    so: $4x + 2 + 12x - 2 = 20x - 12$; giving $x = 3$
    (b) Rect area = 119 units; square area = 144 units;
    $119 \div 144 \times 100\% = 82.6\%$
16. (a) 3 (b) 2 (c) Total goals = $(9 \times 1) + (18 \times 2) + (22 \times 3) + (4 \times 4) + (2 \times 5) = 137$; $137 \div 60 = 2.28$
17.

N	T	Q	Q<160?
16	12	244	NO
15	14	211	NO
14	16	180	NO
13	18	151	YES Answer N = 13

18. (−7, 6)
19. Prime factors of 30 = 2 × 3 × 5; prime factors of 36 = 2 × 2 × 3 × 3; So LCF = 2 × 2 × 3 × 3 × 5 = 180 minutes; time 0720 + 180 minutes = 1020;
20. (a) $(x + 3)(x − 2) = 62 − 12$; giving $x^2 + x − 30 = 0$
    (b) $x^2 + x − 30 = 0$; so $(x + 6)(x − 5) = 0$ giving $x = 5$ or $−6$; but it can't be $−6$ so $x = 5$
21. $(x + p)(x + p) = x^2 + 2px + p^2$; equating the x terms gives $2p = −6$; so $p = −3$; equating the constant terms gives $p^2 + q = 3$ so $9 + q = 3$; giving $q = −6$
22. (a) $(x − 9)(x + 9)$ (b) $(7 − x)(7 + x)$
    (c) $(2x − 5)(2x + 5)$
23. Let lowest mark = x; so range = $24 − x$; upper quartile = 15.5; lower quartile = 8.5; so IQR = 7; we are told that range = 3 × IQR, therefore $24 − x = 3 × 7$; giving $x = 3$
24. (a) $\frac{2x}{3y}$ (b) $\frac{24x^3y^2}{3x^2y} = 8xy$ (c) $\frac{pq}{qr} = \frac{p}{r}$ (d) $\frac{x^2y^3}{x^3y^2} = \frac{y}{x}$
25. Prime factors of 75 (minutes) = 3 × 5 × 5; prime factors of 100 = 2 × 2 × 5 × 5; so LCF = 2 × 2 × 3 × 5 × 5 = 300; 9:20am + 300 minutes = 2:20 pm
26. Perimeter of sector = $2r + 2\pi r × {}^{35}/{}_{360}$; giving $r = 8$ cm; area of sector = $\pi r^2 × {}^{35}/{}_{360} = 19.55$ cm^2
27. (a) $\frac{9x − 6}{18} + \frac{4x + 6}{18} = \frac{13x}{18}$
    (b) $\frac{8x − 6}{4} − \frac{2 − 3x}{4} = \frac{11x − 8}{4}$

Revision Exercise 4

1. (a) $x^2 + 9x − 36 = 0$, giving $x = −12$ or $x = 3$; but it can't be $−12$ so $x = 3$
   (b) 30 cm
2. Gradient = $−8 ÷ 2 = −4$; using $y − y_1 = m(x − x_1)$ gives: $y − 2 = −2(x − (−2))$; $y = −2x$
3. tan 24 = BC ÷ 14; giving BC = 6.23; area of ABCD = 14 × 6.23 = 87.26 cm^2; side of PQRS = $\sqrt{87.26}$ = 9.34 cm; so perimeter PQRS = 4(9.34) = 37.4 cm
4. (a) [Venn diagram: T, P, Y with values 10, 8, 7, 4, 3, 5, 9, 4] (b) 4
5. $y = mx + c$; gradient = 6 as it is parallel so: $y = 6x + c$; substituting $x = 2$ and $y = −4$ gives: $−4 = 12 + c$, so $c = −16$; so answer is $y = 6x − 16$
6. Grad of A = 3; B can be written as $y = −3x + 1½$, so grad B = $−3$; C can be written as $y = 2x + ⅓$, so grad C = 2; grad D = $−3$; so B and D are parallel

7. (a) (b) [two circles, (a) with tangent line, (b) with chord and shaded segment]
8. Pressure = force ÷ area = 310.08 ÷ (7.6 × 4.8) = 8.5 N/m^2
9. Pressure = force ÷ area; so 22 = 36 ÷ area; so area = 1.636 m^2 = πr^2; giving $r = 0.72$ m
10. (a) Circumference of circle = $2\pi r$ = 37.7 cm; 37.7 × 48 ÷ 360 = 5.03 cm (b) 5.03 + 6 + 6 = 17.03 cm (c) $\pi r^2 × 48 ÷ 360 = 15.08$ cm^2
11. Volume = $\pi r^2 h = 60$; curved area = $2\pi rh = 70$; dividing first equation by second gives ½r = 0.8; so $r = 1.6$ cm
12. (a) Length of arc = $36 ÷ 360 × 2\pi r = 9.425$ giving $r = 15$ cm (b) $\pi r^2 × 36 ÷ 360 = 70.69$ cm^2
13. Total number of objects = $14 + x$; Total of all lengths = $(7 × 2) + (4 × 5) + (8 × x) + (3 × 11) = 8x + 67$; since mean length = 5.75 then: $8x + 67 = 5.75 × (14 + x)$; giving $x = 6$
14. (a)

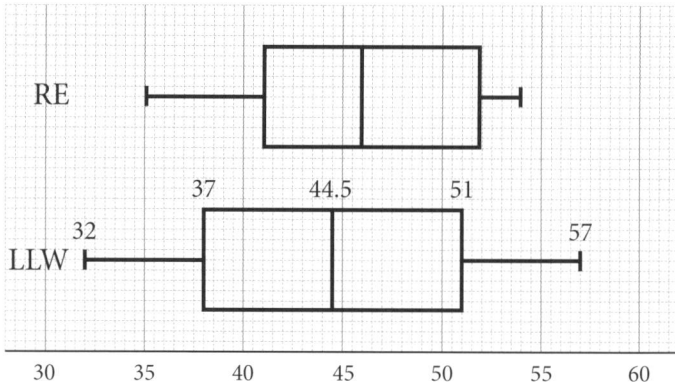

(b) The RE results are more consistent as the marks in LLW are more spread out than the marks in RE. The RE marks are better than the LLW marks as the median RE mark, lower quartile and upper quartile are higher than the LLW marks.
15. (a) Area of closed end = $\pi r^2 = 55.41$; so total curved area = $960 − 55.41 = 904.58$ cm^2
    (b) $904.58 ÷ 2\pi r = 34.28$ cm (c) $\pi r^2 h = 1899.72$ cm^3
16. (a) Total sphere area = $4\pi r^2 = 1017.88$; so area of hemisphere = $(1017.88 ÷ 2) + \pi r^2$ [for base] = 763.41 cm (b) ½ × ⁴⁄₃πr^3 = 1526.81 cm
17. ratio of volumes of A:B = 8:1
    ratio of lengths of A:B = 2:1 (taking cube roots)
    ratio of areas of A:B = 4:1 (by squaring)

Answers

18. Cylinder volume = $\pi r^2 h$ = 19.30 m²; volume of each sphere = $\frac{4}{3}\pi r^3$ = 0.18 m²; 19.30 ÷ 0.18 = 107.2, so 107 spheres can be made
19. (a) $\pi r^2 h$ = 1256.64 cm³ (b) Sloping height = 15.62 cm (by Pythagoras' theorem); surface area = $\pi r l$ = 490.72 cm²
20. (a) Base area = πr^3 = 113.10; so curved area = 720 − 113.10 = 606.9 cm² (b) 606.90 = $\pi r l$ giving l = 32.2 cm; perpendicular height = 31.6 cm (by Pythagoras' theorem) (c) $\frac{1}{3}\pi r^3$ = 1191.29 cm³
21. vol A = 2 × vol B; $\frac{1}{3}\pi a^2 b$ = 2 × $\frac{1}{3}\pi(3a)^2 h$; dividing both sides by $\frac{1}{3}\pi a^2$ gives: b = 2 × 9h, so $h = \dfrac{b}{18}$
22. (a) 16.5 (b) 11 (c) 24 (d) 13
23. Begin by finding two values to give the median, then find values for the two quartiles and finally find the first and last numbers to create the required range. Multiple answers are possible. One example is: 2, 6, 9, 11, 13, 17
24. tan 68 = QT ÷ 24; giving QT = 59.4; tan B = 59.4 ÷ 40; giving B = 56°
25. (a) AD = 14.14 m (by Pythagoras' theorem); area = ½(7 + 12) × 14.14 = 134.35 m² (b) pressure = force ÷ area; giving force = 208 700 N
26. sin ACD = 4.5 ÷ 18 giving ACD = 14.48°; tan ACD = AD ÷ 15 giving AD = 4.65 m; area = 4.65 × 18 = 83.7 m
27. (a) $2\pi r h$ = 2π × 5.5 × 13.5 = 466.5 m (b) $\pi r^2 h$ = π × 6.5² × 14.5 = 1924.6 m²

Meeting the requirements of the two-tier CCEA GCSE Mathematics specification, this is one of eight revision booklets to cover levels M1 to M8. These valuable questions were specially commissioned for the booklet and are not from past papers. Full answers are included at the rear and contain not only the final answer but, where appropriate, an indication of the process required to reach the given solution. The book has been through a meticulous quality assurance process by a GCSE Mathematics expert.

Which revision booklets do I need?

Students sitting CCEA GCSE Mathematics will usually be in one of four pathways and will require two revision booklets. The student's teacher will be able to advise which pathway they are studying.

If the student is studying this pathway...	...they will need these revision booklets
Foundation Tier Option 1	M1 and M5
Foundation Tier Option 2	M2 and M6
Higher Tier Option 1	M3 and M7
Higher Tier Option 2	M4 and M8

COLOURPOINT EDUCATIONAL

£3.99

ISBN 978-1-78073-194-0

www.colourpoint.co.uk